alpha
science

Materials

Nicola Barber

 Evans Brothers Limited

This book is based on *Designs in Science Materials* by Sally and Adrian Morgan, first published by Evans Brothers Limited in 1994, but the original text has been simplified.

Evans Brothers Limited
2A Portman Mansions
Chiltern Street
London W1M 1LE

First published 1997

Printed in Hong Kong

Managing Editor: Su Swallow
Editor: Catherine Bradley
Designer: Neil Sayer
Typesetting: TJ Graphics
Production: Jenny Mulvanny
Illustrations: Hardlines, Charlbury
                David McAllister

British Library Cataloguing in Publication Data

Barber, Nicola
    Materials - (Alpha science)
    1. Materials - Juvenile literature
    I. Title
    620.1'1

    ISBN 023751771X

Acknowledgements

For permission to reproduce copyright material the authors and publishers gratefully acknowledge the following:

Cover David Parker/Seagate Microelectronics Ltd./Science Photo Library (showing a technician in clean-room clothing inspecting silicon wafers).
Title page Roberto de Gugliemo, Science Photo Library
Contents page Sally Morgan, Ecoscene
page 4 Michael Holford **page 5** (left) Douglas Dickens, NHPA (right) Richard Leeney, Ecoscene **page 7** Tweedie, Ecoscene **page 8** (left) Merrell Wood, The Image Bank (right) David Overcash, Bruce Coleman Limited **page 9** Roberto de Gugliemo, Science Photo Library **page 10** David Taylor, Science Photo Library **page 11** (inset) Chinch Gryniewicz, Ecoscene (below) Robert Harding Picture Library **page 12** Sally Morgan, Ecoscene **page 13** Robert Harding Picture Library **page 14** (top) Sally Morgan, Ecoscene (below) J.C. Revy, Science Photo Library **page 15** Martin Chillmaid, Oxford Scientific Films **page 16** (top left) Stephen Dalton, NHPA (inset) Sally Morgan, Ecoscene (below left) J. C. Revy, Science Photo Library **page 17** (top) Robert Harding Picture library (bottom) Sally Morgan, Ecoscene **page 18** (left) Lees, Ecoscene (inset) Eric Soder, NHPA **page 19** Sally Morgan, Ecoscene **page 20** Anthony Cooper, Ecoscene **page 21** (top) Sally Morgan, Ecoscene (bottom) Astrid and Hanns-Frieder Michler, Science Photo Library **page 22** (inset) David Woodfall, NHPA (bottom) Martin Wendler, NHPA **page 23** Sally Morgan, Ecoscene **page 24** (left) Sally Morgan, Ecoscene (main picture, bottom right) Robert Harding Picture Library **page 25** Sally Morgan, Ecoscene **page 26** (left) William S. Paton, Planet Earth Pictures (inset) Winkley, Ecoscene **page 27** Astrid and Hanns-Frieder Michler, Science Photo Library **page 28** Sally Morgan, Ecoscene **page 29** Robert Harding Picture Library **page 30** (top) Adrian Davies, Bruce Coleman Limited (bottom) Walter Murray, NHPA **page 31** (top) B. Kloske, Ecoscene (bottom) A.N.T., NHPA, (inset) Mark Caney, Ecoscene **page 32** (top) Robert Harding Picture Library (bottom) Ayres, Ecoscene **page 33** Prof. P. Motta, Dept. of Anatomy/University of 'La Sapienza', Rome/Science Photo Library **page 34** (top) Steve Turner, Oxford Scientific Films (inset) A.N.T., NHPA **page 36** Tim Thackrah **page 37** Sally Morgan, Ecoscene **page 38** (left) Sally Morgan, Ecoscene (inset) Lees, Ecoscene **page 39** Sally Morgan, Ecoscene **page 40** (top) Roy Walker, NHPA (bottom) Sally Morgan, Ecoscene **page 41** (top) Steve Hopkin, Planet Earth Pictures (bottom) Sally Morgan, Ecoscene **page 42** Lawrence Livermore National Laboratory, Science Photo Library **page 43** (top) G.I. Bernard, NHPA (bottom) Fritz Prenzel, Bruce Coleman Limited

Contents

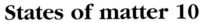

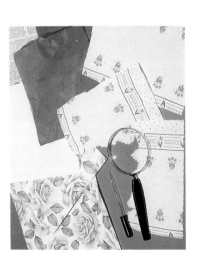

Introduction

The word material is used to describe what something is made of. We talk about the kind of material used to make clothes. We talk about building materials to describe bricks, or concrete, or stone.

There are thousands of different kinds of materials. Some of these materials are natural, such as cotton or wood. Other materials are made by people, such as plastic or glass. Some materials are very hard or very strong. Others are soft, or break very easily.

The first materials used by people were wood, clay and rock. Early people also found metals such as gold in the ground. They hammered the gold to work it into shape. Later, people learned how to change materials by heating them, or by mixing them with other substances.

When people plan to make something, they need to choose their materials very carefully. Does the material need to bend? Does it need to be hard, or soft? Does it need to be light or heavy? A designer needs to think about all these questions.

In this book you can find out about many different kinds of material. In each section you will find some amazing facts, some experiments for you to try, and

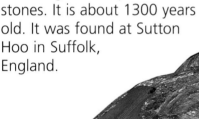

This helmet is made from leather, gold and precious stones. It is about 1300 years old. It was found at Sutton Hoo in Suffolk, England.

▽ Beavers use natural materials such as wood to build dams across streams.

▷ Modern buildings are made from many different kinds of materials, some natural and some man-made.

Measurement
In this book, some measurement words are shortened:

kilometre	km
metre	m
centimetre	cm
cubic centimetre	cm³
kilogram	kg
gram	g
degrees Celsius	°C
(to measure temperature)	

some questions for you to think about. At the end of each section, you will find a box called **Key words**. These boxes explain important words in the text. You can also look up difficult words in the Glossary on page 44.

Key words
Material the substance from which something is made.
Designer a person who plans the shape and style of something.

Atoms and elements

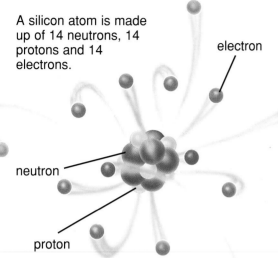

A silicon atom is made up of 14 neutrons, 14 protons and 14 electrons.

electron

neutron

proton

! *An atom is so tiny that four billion sodium atoms would fit on the full stop at the end of this sentence.*

Materials are made from particles called atoms. Atoms are very, very tiny. They are so small that scientists cannot see them even with the most powerful microscopes. However, scientists know quite a lot about atoms.

Atoms are made of three kinds of smaller particles. These are called protons, neutrons and electrons. The protons and neutrons make up the middle of the atom. This middle part is called the nucleus. The electrons move about in layers around the nucleus. The smallest atom is hydrogen which has one electron. The largest atom is uranium which has 92 electrons.

Atoms are found on their own, and in groups. A group of atoms is called a molecule.

Elements

An element is a substance made up of only one kind of atom. Copper is an element because it is made up only of copper atoms. There are 92 different natural elements. There are 12 more elements made by people. These are called artificial elements.

An element cannot be broken up to make a different substance. For example, oxygen is an element made up of oxygen atoms. It cannot be broken down into a smaller or simpler element. Elements are the 'building blocks' that make up all materials.

Different elements can join together. For example, water is made up of two different elements, oxygen and hydrogen. A material

! *Most of your body is made up of just four elements – oxygen, hydrogen, carbon and nitrogen.*

The seaweeds and sea animals in this rockpool are made up of many different compounds.

such as wood is made up of many different elements. When different elements are joined together, this is called a compound.

Chemical reactions

You can change a substance by heating it, by burning it, by adding water to it, or by mixing it with another substance. This process is called a chemical reaction. Once a chemical reaction has happened, it is difficult to undo. For example, a chemical reaction happens when you boil an egg. As you heat the egg, the white and yolk become firmer. Once the egg is boiled, you cannot make it into a raw egg again.

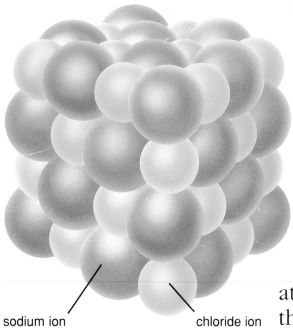

sodium ion   chloride ion

The atomic structure of salt (sodium chloride). The atoms in salt form a regular pattern held together by bonds.

What happens in a chemical reaction? The atoms in one substance react with the atoms in another substance. The atoms move towards each other to form new bonds. A bond is an attraction between two atoms. It's a bit like gluing two atoms together. The new arrangement of atoms forms a new substance. For example, fuels such as coal, oil and gas contain the element carbon. When you burn carbon in air, the atoms in the carbon react with atoms in the air. The atoms form new bonds to make a compound called carbon dioxide.

Sometimes the bonds between atoms are broken. For example, a kind of sugar called glucose is made up of carbon, hydrogen and oxygen atoms. When the bonds between the atoms in glucose are broken, two smaller

Melting metal is a physical reaction. Although the bonds between the atoms are broken, they will re-form when the metal cools and turns into a solid again.

Fireworks make use of chemical reactions. The chemicals in the fireworks react as they burn in the air.

compounds are made. These compounds are water and carbon dioxide.

Physical reactions

In a chemical reaction, atoms react to form new bonds. In a physical reaction, atoms do not react. An example of a physical reaction is when sugar dissolves in water. The atoms in the sugar and the water do not react with each other. This means that it is possible to separate the sugar from the water again.

Atomic structures

Some materials are made up of simple molecules. Others have a more complicated structure. Sand is a very common material, but it is has a very complicated arrangement of atoms. Sand is made from a compound called silicon dioxide. Silicon dioxide is made when atoms of silicon bond with atoms of oxygen. These atoms bond together firmly to make a strong structure. This makes sand a tough substance that is often used in building materials.

In metals, the atoms are packed tightly together in a regular pattern. They are held firmly in position by bonds. If you want to melt a metal, you have to heat it to a very high temperature. This is because it takes a lot of heat energy to break the bonds between the atoms in a metal. However, the bonds in a metal are quite flexible.

A crystal of quartz. Quartz is made from silicon dioxide.

Key words
Atom the smallest part of a chemical element that can exist alone.
Bond an attraction between two atoms.
Compound a substance made from two or more elements.
Element a substance made up of only one kind of atom.
Molecule a group of atoms bonded together.

There is a very rare state of matter called plasma. This state happens when gas is heated to a very high temperature, making the atoms break apart. The Sun is a ball of plasma.

States of matter

Materials can exist as solids, liquids or gases. Wood, metal and ice are all examples of solids. Water is a liquid. The air you breathe is a gas. Materials can change from one state to another. For example, when water freezes into ice, it changes from a liquid into a solid. When it melts, it changes from a solid back into a liquid again.

Inside a solid, the atoms are packed closely together and held in place by strong bonds. This means that a solid has a fixed shape. Inside a liquid, the atoms move about. This means that liquids can flow easily and change shape. In a gas, the atoms are not bonded together at all. If you release a small amount of gas in a room, the gas atoms will spread out across the room.

The air inside a hot-air balloon is heated by a burner. Hot air is lighter than cold air, so the balloon rises up into the sky.

Using solids, liquids and gases

Solids are hard materials that hold their shape. This makes them very useful. Look around you and you will see examples of solids such as bricks, wood, paper, plastic, metal, concrete, and many more.

Liquids are often used as lubricants. A lubricant is a slippery material that is used to make two surfaces slide over each other easily. For example, oil is used as a lubricant in a car engine. Without any oil, the metal parts in an engine would grind against each

Why are salt and grit spread on roads when ice or snow is expected?

other and quickly wear away.

Gas can be squashed into a small space. Car tyres contain compressed (squashed) air. Divers use containers full of compressed air to breathe under water. They carry these containers on their backs. Gas can also be heated to make it expand. The air inside a hot-air balloon is heated to make it lighter than the colder air surrounding the balloon. This is why hot air balloons float in the sky.

Changing state

What happens when a solid is heated? The heat energy makes the atoms in the solid move. If there is enough energy the atoms break their bonds. Then the solid melts and turns into a liquid. If you continue to heat the liquid, the atoms move even faster. Eventually they break away from the other atoms in the liquid. Then the liquid evaporates and becomes a gas.

Many materials can exist as a solid, a liquid and a gas.

When a volcano erupts, melted rock called lava pours out of the volcano (big picture). When the lava cools it turns into solid rock (small picture).

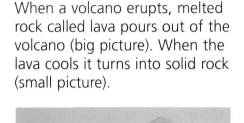

EXPERIMENT

States of matter

In this experiment you can find out about how atoms move in different states.

You will need: a square lid from a box, some marbles, some pieces of cardboard, some sticky tape or glue.

1. Divide the lid into three sections as shown in the photograph, using thin strips of cardboard. Fix the cardboard into place with the sticky tape or glue.

2. Fill the small space with marbles. Count the number of marbles you have used.

3. Now put the same number of marbles in each of the two other spaces.

4. Gently move the lid around.

The closely packed marbles cannot move very much. This is like the atoms in a solid. The

marbles in the next section can move more. This is like the atoms in a liquid. The marbles in the largest section can move a long way. This is like the atoms in a gas.

Why do people use 'dry ice' for special effects in films, or in the theatre?

Some materials never exist as liquids. Instead they turn directly from a solid into a gas. This is called sublimation. For example, a solid called iodine turns directly into a purple gas when it is heated. The same process can happen in reverse. Carbon dioxide gas turns directly into a solid when it is cooled to -55°C. This solid is called 'dry ice', often used for special effects in the theatre.

Gels

A gel is a special state of matter between a liquid and a solid. It is made by mixing a liquid and a solid. The solid makes the liquid flow less easily. The liquid makes the solid less firm. If you squeeze a gel gently, it will change shape. But it will go back to its original shape when you stop squeezing. If you squeeze a gel with more force, it will flow like a liquid.

You can see the shiny mucus produced by this snail to help it move along the ground.

A natural kind of gel is mucus. Fish are covered in a layer of slippery mucus to help them swim through the water. Snails and slugs produce mucus to help them move along the ground.

Grease and Vaseline are gels. They are used as lubricants (see page 10) to stop moving parts from rubbing.

Many 'non-drip' paints are gels. Gels are also found in food. Jelly contains a gel called gelatine. Jams contain a natural gel called pectin. Pectin makes jam become firm as it cools. Jam is solid enough to stick to a spoon, but runny enough for you to spread.

Key words
Gas a substance made up of atoms that can move around freely.
Gel a substance that is part solid and part liquid.
Liquid a substance made up of atoms that can move a little.
Solid a substance in which the atoms are held together by strong bonds.

EXPERIMENT

Making a gel

Flour is often used in cookery to thicken sauces and soups. Flour contains starch which forms a gel when heated in water. You can make your own gel with flour and water.
You will need: 200g of flour, 100 cm^3 of water, a saucepan, a wooden spoon, a tablespoon. You need to use a cooker, so you may need to ask an adult to help you with this experiment.
1. Put one heaped tablespoon of flour in the saucepan. Add the cold water slowly, stirring all the time so that the flour mixes well with the water.
2. Heat the mixture over a low heat, stirring all the time.
3. Remove the saucepan from the heat when the mixture starts to bubble. How thick is the gel? Allow it to cool. What happens as the gel cools?
You can try this experiment again with more flour.

Polymers

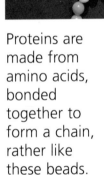

Proteins are made from amino acids, bonded together to form a chain, rather like these beads.

This plant cell is full of starch grains.

Polymers are large molecules. Each polymer molecule is made up of many small parts called monomers. The monomers form a long chain. Polymers always contain carbon, hydrogen and oxygen. They are found in every living thing on Earth. There are many different kinds of polymers, both natural and man-made.

There are two important groups of natural polymers, proteins and carbohydrates. Protein is a natural polymer found in living things. It is made from monomers called amino acids. The amino acids bond together in long chains. The human body contains protein molecules made from 20 different amino acids. Proteins are used in many different ways in the body, for example in the muscles, and inside cells.

Carbohydrate polymers include starch and cellulose. Starch is made of glucose, a kind of sugar. Plants need supplies of starch to survive. They store the starch in their leaves and roots for food. Cellulose is also a polymer made from glucose. It is used to make the cell walls around every plant cell. Cellulose is a tough material. It gives a cell its shape, and helps to protect the inside of the cell.

People learned to make man-made polymers about 100 years ago. At first, these polymers were very expensive to make. But in the 1950s, new supplies of oil were found. Most man-made polymers are now made from oil. Man-made polymers are found in plastics such as polystyrene and polyvinylchloride (PVC). Today, they can be made very cheaply.

There are two main groups of polymers. These are called fibres and plastics.

Fibres

A fibre is a long, thin thread of material. Some fibres come from plants, such as cotton or the fibres in paper. Some fibres, such as wool, come from animals. There are also many man-made fibres such as nylon or rayon. Sometimes, fibres are twisted together to make a long thread. For example, wool fibres are twisted together to make thread for weaving cloth.

Your body contains many different kinds of fibre, and each has its own job. Your hair and nails are made from a kind of fibrous protein called keratin. All animal hair, nails, hooves and horn are made from keratin. In keratin, the chain of amino acids is twisted, like the cord of a telephone. This twisted chain is strong and elastic. You can stretch your wet hair quite a lot and it will still spring back into shape.

Animal hair is made from a kind of fibrous protein called keratin.

What happens to woollen clothes if they are washed in a very hot wash? Why?

Silk

Silk is a natural fibre. Many animals produce silk, including spiders when they spin their webs. But the silk used by people to make cloth comes from the silkworm moth.

The silk is actually made by the caterpillar

It takes silk from more than 3000 silkworm cocoons to make the finest kimonos.

◁ The silk moth caterpillar spins a cocoon of silk.

△ Silk has a special feel and look that cannot be matched by man-made fibres.

of the silkworm moth. The caterpillar spins a long thread of silk around its own body. This thread makes a protective layer called a cocoon. A caterpillar can spin one unbroken thread that is up to 2 km long! To harvest the silk, the farmer puts the cocoon into water. The water loosens the thread. The farmer then unwinds the silk.

Plant fibres

A close-up of fibrous material in a plant.

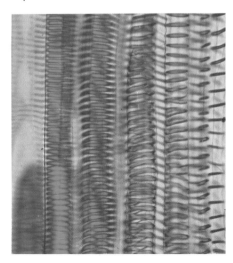

Plants make a lot of fibrous material. Some plants make strong fibres that are difficult to break.

People use the fibres in plant stems to make many materials. Flax, sisal, jute and hemp are all materials made from plant stems. The farmer cuts the plant stems and puts them in water. The outer covering of the plant stem is removed, and the fibres are separated out. Then these tough fibres are twisted together to make string or rope, or woven together to

Harvesting cotton. You can see the round cotton bolls on the plants.

make cloth. People use flax fibres to make a material called linen. Sisal fibres are used to make string, rope, nets and carpet.

One of the most important natural fibres is cotton. Cotton fibres are wrapped around the seeds of the cotton plant in a ball called a 'boll'. The cotton boll is picked off the plant. The seeds and the cotton fibres are separated. A machine untangles the cotton fibres. Then the fibres are spun into one long cotton thread.

Cotton is not only used for making clothes. How many other uses can you think of for cotton?

Man-made fibres

Another word for man-made is 'synthetic'. There are many synthetic fibres including rayon, nylon, Kevlar and Lycra. Synthetic fibres are usually stronger than natural fibres.

Rayon is made from the plant material cellulose (see page 14). Small pieces of wood are put into acid. The acid turns the wood into pulp. This pulp is washed and coloured white with bleach. Then it

A long sheet of cellulose on a roll. Cellulose comes from plant material.

is rolled into a long sheet. Cellulose has many different uses. It is used to make rayon and viscose cloth. It can be stretched out to make fine threads to be used in teabags or disposable nappies. It is used in Sellotape, in face creams and even in instant mashed potato! Cellulose is also used to make cellophane.

Nylon is made from oil. It was the first synthetic fibre to be made. Nylon is strong and light. Unlike cotton and wool it does not shrink in hot water or rot. There are hundreds of different kinds of nylon. Tights and stockings are made from nylon. Kevlar is a particularly tough kind of nylon. It cannot be cut with scissors. It is often used to make bullet-proof vests.

Materials such as Lycra and Elastane are more recent inventions. These materials stretch and are often used in sports clothes.

Many materials need to be waterproof. The best waterproof clothes keep the rain out but

People wear waterproof clothes to keep water out.

Birds waterproof themselves by spreading a kind of oil over their feathers.

EXPERIMENT

Testing fabrics

When designers choose a material for clothes or for furniture, they have to think carefully about how the material will be used. In this experiment you can test some materials to see how waterproof they are, and how well they stand up to being rubbed.

You will need: some different materials, for example cotton, wool, nylon, viscose, PVC; a jam jar, an elastic band, a measuring jug, a rough stone or piece of sandpaper.

1. Cut out a small square of material and stretch it across the top of the jam jar. Fix it in place with the elastic band.
2. Test the material for waterproofing by dripping water on to the surface of the material. Does the water soak through or stay on top of the material?

3. Rub the material with the stone or sandpaper. How long do you have to rub before the material wears through?
4. Try the experiment again with the other materials.

 Look at your clothes. How many different materials are you wearing?

allow water vapour from the skin to escape. Many animals waterproof their own fur or feathers. Birds produce a kind of oil which they rub over their feathers, using their head and feet. The oil keeps water out, and helps to protect the feathers.

Even human skin is waterproof. Your skin is made from many layers of cells. The outside layer is made from dead cells which contain keratin (see page 15). The keratin stops water getting into the skin. There is a layer of an oily substance called sebum on the surface of the skin. This also helps waterproof the skin.

Plastics

 A modern car contains at least 100kg of plastic.

We use plastics more and more in our everyday lives. Plastics are used to make toys, buildings, computers, kitchen equipment, furniture, clothes – the list is endless. Plastics are polymers (see page 14) made from oil. They are cheap, light, and easy to shape and colour.

There are two main groups of plastics. They are called thermoplastics and thermosetting plastics. Thermoplastics can be melted by heating. This means that old thermoplastics can be melted down and used again to make new objects. Thermosetting plastics do not soften when they are heated. Instead, they stay hard and in shape. If they are heated to a high temperature they will burn. Thermosetting plastics are used to make objects such as electrical fittings that must not melt when they get hot.

Some polymers are used both as fibres and as plastics. For example, nylon (see page 18) is used as a fibre to make materials and toothbrushes, and as a plastic to make objects such as gear wheels.

In our homes we use many objects made from plastic.

Rotting materials

Plastics are cheap and useful, but they do have one disadvantage. They don't rot away when people have finished using them. They are not biodegradable. Natural polymers such as cotton or cellulose will quickly decay if they

EXPERIMENT

Biodegradable materials

In this experiment you will bury some materials in the ground and then dig them up again a few weeks later. You will find out which materials are biodegradable and which are not. You will need: four different materials, for example newspaper, cabbage leaves, a plastic cup and an eggshell; a knife, scissors, a spade, some sticks or canes, a pair of plastic gloves, a few labels and a marker pen.

1. Collect your four materials.
2. Go into your garden and dig four holes in the soil. The holes should be close together and about 10 cm deep. Put each material in a different hole and cover them with soil.
3. Mark the place where you have buried each material with a stick. Write on a label what is buried in each hole, and tie the label to the stick.

4. Leave the materials in the soil for two weeks. If you do this experiment in winter, leave the materials for at least one month.
5. Put on the plastic gloves and dig up the materials again. Which ones have broken down?

are left outside. This is because animals and plants called bacteria and fungi break down and digest the natural polymers. However, a piece of plastic left outside will remain the same for many years.

Scientists have now learned how to make plastics that will break down. These new plastics are called biodegradable plastics.

Why do biodegradable materials take longer to break down in winter?

Many waste dumps are full of old plastic that will never rot away

Key words
Biodegradable able to be broken down.
Fibre a long, thin thread of material.
Plastic a man-made material made from oil.
Polymer a large molecule made up of smaller parts.

Building materials

Many different materials are used for building structures, both natural and man-made. The best building materials are strong and long-lasting.

Wood

People have used wood to make shelters for thousands of years. Wood is also useful for many other things, for example making fences or furniture. There are two main groups of wood: hardwoods and softwoods.

Hardwood comes from trees such as the oak or the beech. These trees have a thick, hard wood which is very long lasting. Hardwood rots very slowly, so it is a good wood for building. Hardwoods such as mahogany and teak come from tropical rainforests.

Hardwood trees such as mahogany and teak grow in the tropical rainforests (above). People have used these woods for hundreds of years to make high-quality furniture. Loggers cut the trees down, and take them to the sawmill (right) to be cut into planks.

New wood contains too much water so it is left outside to dry. What happens to the water?

Softwood comes from conifer trees such as the pine and spruce. The wood from these trees is softer and not as tough as the wood from hardwood trees.

Wood is a renewable resource. This means that new trees can be planted to replace the trees that are cut down. Softwood trees grow quite quickly. They can be used between 10 and 20 years after they were planted. However, hardwood trees take a long time to grow.

In the USA, Canada and northern Europe, there are large forests of softwood trees. Softwood trees are an important crop for these countries. When the trees are cut down they are taken to a sawmill. Here, the trunks are cut up. Some of the wood is used to make plywood. Plywood is made from several thin sheets of wood glued together. It is a very strong material.

None of the wood is wasted. Small pieces of wood are used to make paper (see page 24) or cellophane (see page 18). Even the oils from the wood can be used to make various products such as tar, creosote and turpentine.

The trunks of softwood trees are often taken to the sawmill by floating them down a river.

Paper

People make paper and cardboard from wood pulp. Wood pulp is full of cellulose (see page 18).

The quality of the paper depends on the length of the cellulose fibres in the pulp. Paper made from long fibres is very strong. New paper can also be made from old, recycled paper. However, the fibres in recycled paper are shorter, and the quality is not usually as good.

Some animals make paper too. Wasps make their nests from paper. They produce a paste by chewing wood until it becomes a pulp. They put the paste in place with their forelegs. The paste dries in the air.

! *It takes 40,000 litres of water to make one tonne of paper!*

△ In the first stage of making paper, wood pulp is mixed with water and other substances.

△ Water drains through holes in the conveyor belt.

△ After passing through ovens, the dry paper is wound on to rollers.

△ The finished paper in large rolls

? *Can you think of six different ways that people use paper?*

Wasps make their nests from a kind of paper.

EXPERIMENT

Examining paper

The strength of paper depends on the length of fibres used to make it. In this experiment you can test the strength of different 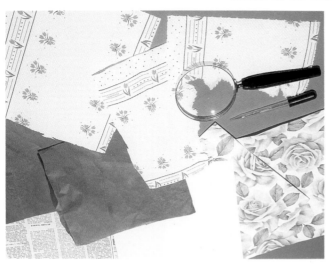 kinds of paper, and the amount of water they will absorb.

You will need: a selection of different papers, for example recycled paper, writing paper, newspaper, paper towels; an eye dropper, coloured water or ink, scissors, a magnifying glass.

1. Cut your pieces of paper into squares measuring about 20 x 20 cm. You will need two squares of each kind of paper.

2. Test each kind of paper for strength. How easy is it to tear? Use the magnifying glass to look at the fibres along the torn edge of the paper.

3. Now test each kind of paper to see how much water it will absorb. Fill the eye dropper with coloured water or ink. Drip a few drops on to the first piece of paper. Watch what happens. Does the water sit on top of the paper or does the paper absorb it? Try this experiment with the other pieces of paper.

Clay

People and animals use clay to build their homes. Clay is made from finely ground rock. When the clay is wet, it is easy to shape. As the clay dries it hardens to become a solid.

In Africa, Central America and the south of the USA, people make their houses with clay. They use clay bricks or wood to make a framework. Then they cover the framework with layers of clay. Each layer dries in the sun before the next layer is put on. In Mexico and the USA this kind of house is called an adobe house.

In places where there are hot days and cool nights, clay houses help to keep a regular temperature inside the house. The clay walls

People made the first clay bricks about 8000 years ago in Mesopotamia (present-day Iraq).

absorb heat during the day. This keeps the inside of the house cool compared to outside. At night, the clay walls release the heat. This warms the inside of the house compared to the temperature outside.

Many birds use clay to make their nests. House martins (above) build clay nests under the eaves of houses.

An adobe house made of clay.

Metals

Of the 92 natural elements (see page 6), 81 are metals. Metals are strong, hard materials which bend and stretch easily (see page 9). Sometimes a metal is used on its own but often, metals are mixed together to make them stronger and harder. Brass is a mixture of two metals, copper and zinc. A mixture of metals is called an alloy.

Gold is sometimes found in lumps called nuggets in the ground!

Zinc is a blue-white metal.

A blast furnace

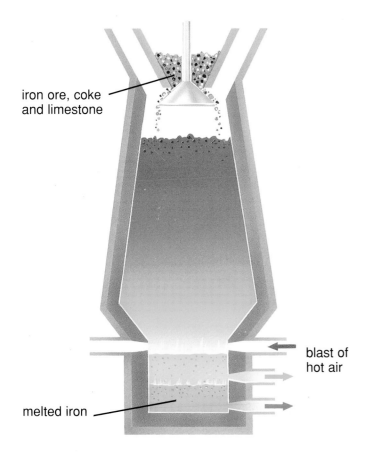

iron ore, coke
and limestone

blast of
hot air

melted iron

Scientists describe most metals as reactive. This means that the atoms in most metals react easily with atoms in other substances. The only metals that do not react easily are gold and platinum.

Metals are usually found in compounds called ores. An ore is a mixture of metal and rock. People take ore out of the ground in quarries and mines. Then they have to separate the rock and the metal. This is known as extracting the metal.

Iron is one of the cheapest and most important metals. It is extracted from iron ore by a process called smelting. This happens in a blast furnace (see diagram). Iron ore, coke (a kind of coal) and limestone are put into the top of the blast furnace. Hot air blows in at the bottom. Several chemical reactions take place before the iron separates from the ore. The iron melts and turns into a liquid. The melted iron comes out of the bottom of the furnace. This kind of iron is called pig iron. This smelting process is also used to make steel, a very tough metal.

EXPERIMENT

Examining corrosion

Corrosion is a chemical reaction between a metal and another chemical. The most common reaction to cause corrosion is between a metal and oxygen (in water or in air). A new chemical substance forms on the surface of the metal. When iron corrodes, rust forms on its surface. The scientific name for rust is iron oxide. In this experiment you can see how quickly iron rusts in different conditions.

You will need: six small jam jars with lids, six shiny iron nails, a kettle, some salt, some oil.

1. Put one nail in a dry jar and replace the lid.
2. Fill the second jar two-thirds full of tap water. Put a nail in and replace the lid.
3. Boil some water in a kettle. This gets rid of any air in the water. Let the water cool and then fill the third jar two-thirds full. Put a nail into the water. Then add a layer of oil on top of the water. The nail in this jar is now in oxygen-free water. Replace the lid.
4. Put a little water in the fourth jar. Place the nail so that it is half in and half out of the water.

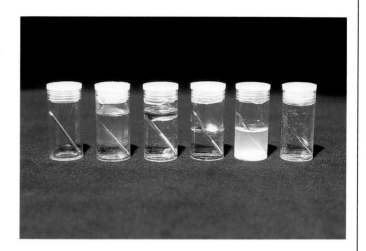

5. Put some water in the fifth jar and add some salt. Put the nail in and replace the lid.
6. Fill the sixth jar with tapwater. Dip the nail in some oil. Put the oil-covered nail in the water. Replace the lid.
7. Leave all the jars for at least a week. Then examine each nail. Which nails have gone rusty? Can you explain why?

Glass

People have made and used glass for about 5000 years. Glass is quite tough and strong, and yet it is transparent (you can see through it). It is made from sand (silicon dioxide), soda lime and limestone. When these substances are heated to 800°C they melt to form glass. At this stage the melted glass is shaped into sheets or into bottles and other objects. Then it is cooled.

The flat sheets of glass used for windows are made by a special method. The melted glass is poured on to a bath of melted tin. It is left to cool slowly. This means that the glass is absolutely flat and even. The glass is then cut into lengths.

! *Old glass in windows is often thicker at the bottom than at the top. This is because glass is really a kind of liquid and the molecules in the glass have moved very slowly downwards.*

A glass-blower shapes melted glass by blowing air into it to make a bottle or a glass.

Expansion and contraction

One problem with many building materials is expansion and contraction. Many materials expand (get bigger) when they get hot. This is because heat energy makes the molecules in the material move faster. As they move they take up more space. The molecules get farther apart and the material expands. The opposite happens when a material gets cold. The molecules move less quickly and take up less space. The material contracts (gets smaller). The change in size is usually very small. But it can make a difference in a large structure such as a bridge or a runway. To avoid this problem, concrete and metal structures are built with small gaps. This allows the material to expand safely.

! *Pyrex is a kind of glass that hardly expands when it is heated. This makes it a good material for cooking pots.*

Key words
Corrosion a chemical reaction between air, water or another chemical on the surface of a metal.

Calcium-rich materials

Calcium is a very common element. It is found in rocks and in living things. Calcium is often combined with other substances such as carbonates (a salt containing carbon and oxygen). Calcium is an important material in both natural and man-made structures.

Calcium carbonate is found in chalk and limestone rocks. People dig it out of the ground for use in many industries. For example, it is used to build roads, and to make glass (see page 28). However, calcium carbonate reacts very easily with acids. (The word acid comes from a Latin word meaning 'sour'.) Rainwater is slightly acid, and it reacts with limestone rock, wearing the rock away. This is why you find many caves in areas with chalk and limestone rocks.

In some places, rainwater has become more acid than usual because of industrial air pollution. Waste gases in the air react with the rainwater to form sulphuric acid. The sulphuric acid falls as rain and attacks limestone buildings. The rain eats away at the stone of the buildings.

Many buildings and statues are made from limestone rock. Acid rain attacks this rock (above), wearing the stone away. These white chalk cliffs are also worn away by rainwater.

Pipi shells, like all other shells, are made from calcium carbonate in layers of protein.

Shells and skeletons

Many animals that live in the sea cannot move about. They have soft bodies which they protect by building hard tubes or shells. These tubes or shells are made from calcium carbonate.

Mollusc shells come in many different shapes and sizes. There are spiral shells, dome-shaped shells, and the huge hinged shells of the giant clam. These shells contain calcium carbonate in layers of protein (see page 14). As the animal grows inside, it adds new layers of shell.

Coral animals build a skeleton to support and protect their soft bodies. They live in large groups, or colonies. Their skeletons are made from calcium carbonate. When an animal dies, its coral skeleton is left behind. In time, the coral gets larger and larger as other animals build their skeletons on the old ones.

Sponges (above) and coral animals (right) make a calcium carbonate skeleton to support themselves.

Cement and concrete

How many different uses of concrete do you know?

Cement and concrete are both man-made materials. They are made from common materials such as limestone, clay, sand and gravel.

People make cement by mixing powdered chalk or limestone with clay and water. The mixture is heated in a kiln. This forms a substance called clinker. The clinker is mixed with gypsum or calcium sulphate and crushed. This is cement in its dry form. To use the dry cement, people mix it with water and sand.

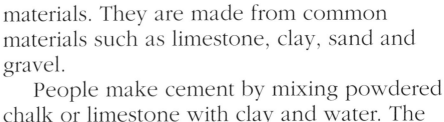

▽ Concrete is made by mixing sand, gravel and cement with water.

Cement breaks quite easily, so scientists are trying to find ways of making it stronger. They have studied cement through microscopes. They can see that cement is full of air holes. If they could get rid of these air holes, cement would be a stronger material. This can be done partly by mixing the cement more carefully, so that less air enters the mixture. It is also possible to add chemicals to make the cement mix better. These improvements mean that cement may be used for more jobs in the future.

Concrete is a strong material. It is useful for building roads and bridges. It is made by mixing sand or gravel

▷ To make reinforced concrete, the concrete is poured over a network of steel rods.

There is a natural kind of cement in your mouth. Your teeth are held in place in your jaw by fibres and a kind of cement.

with cement and water. The strength of concrete comes from the large particles of sand and gravel. People sometimes make concrete even stronger by adding a skeleton of metal, such as steel. The liquid concrete is poured over a network of steel rods. This is called reinforced concrete.

Bones

Fish, amphibians (such as frogs), reptiles (such as lizards), birds and mammals (such as humans) all belong to the family of vertebrates. This is because they all have skeletons inside their bodies. These skeletons are made from two tough materials, cartilage and bone. Cartilage is a fibrous material that can twist and stretch. Bone is stronger and harder. The skeleton of a shark is made entirely from cartilage. The skeleton of a human is made from a mixture of bone and cartilage.

Bone is made from several different substances. Most of the bone is made from a white material which contains the minerals calcium and phospate to make it hard and strong. It also contains collagen, a kind of protein (see page 14). These fibres of collagen make the bone slightly elastic. This helps to stop the bone breaking.

Blood vessels, nerves and living bone cells run through the bone. The living bone cells make the white material. They also repair the bone if it is broken.

Each bone in your body has a different job to do. Some bones need to be very strong. Others need to be very hard, but not particularly strong. The amount of minerals

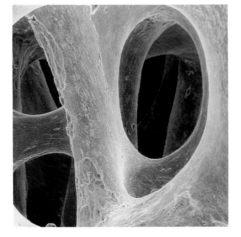

Bone is made from calcium phosphate and tough fibres of collagen.

Bone is as strong as reinforced concrete.

Elephants' ivory tusks (above) are tough and elastic so that they do not break during a fight.

A foal has stiff, strong bones when it is born.

!
An elephant's ivory tusks can grow to be several metres long and weigh more than 45kg.

(calcium and phosphate) and collagen fibres in the bone changes depending on the job of the bone.

In your ear there are some tiny bones called ossicles. These bones carry sound energy from outside deep inside the ear. These bones need to be very stiff, and they must not bend. They are well protected by the thick bone of the skull, so they do not need to stretch or twist. These bones contain a lot of minerals to make the bone very hard.

In contrast, a bone such as a deer's antler has to be very strong. The male deer uses its antlers to fight with other males. The antlers need to be strong and elastic so that they will not break easily. Antlers have a lower amount of minerals than ossicle bones, but they have a higher amount of collagen protein.

There is a difference between the bones of human adults and children. Children's bones are more elastic than adults' bones. As children get older, the amount of mineral in their bones gradually increases. This means that their bones become stronger. However, not all young animals have elastic bones. Horses and antelopes need to be able to get up and run as soon as they are born. These young animals have bones which contain more minerals, and which are quite stiff and strong.

Fibreglass

Fibreglass is a man-made material very similar to bone. This material is made from stiff glass fibres and plastic. The glass fibres are tough and strong. The plastic is stretchy and elastic. Fibreglass is a material that can be bent without breaking. A good example of how fibreglass is used is in a polevaulter's pole. The pole will bend almost double, yet it is very strong and very light.

Teeth

Your teeth are designed for biting and chewing. Each tooth is made up of different layers. The layer on the outside is enamel. Enamel is one of the hardest known natural substances. It is mostly made from calcium phosphate. Beneath the enamel is a layer of dentine. Dentine contains fibres of protein as well as calcium phosphate. This makes it softer

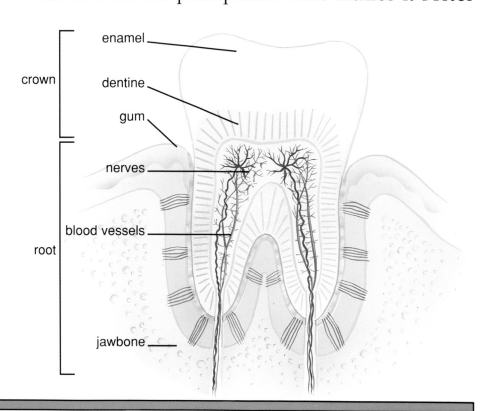

The tooth is in two main parts. The crown is above the gums. The root is in the jawbone.

EXPERIMENT

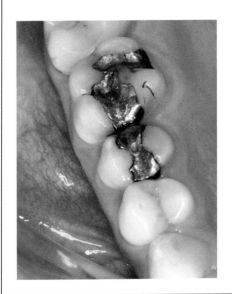

Examining teeth

You will need a small mirror, a notebook and pencil, some friends and family.

1. Use the mirror to help you examine your own teeth. How many teeth do you have? An adult has a full set of 32 teeth, but children do not get their full set until their late teens or 20s.

2. Do you have any fillings? Are they mercury amalgam fillings (see page 37), or a different kind?

3. Examine the teeth of your friends. Do they have any fillings?

4. Examine the teeth of a parent or older member of your family. Does he or she have 32 teeth? Do older people have more fillings than younger people?

 Why is it important to clean your teeth after a meal?

and more elastic than enamel.

It is very important to look after your teeth because they can rot and decay. If acid attacks the enamel layer on your teeth, the enamel starts to rot away. Sometimes the decay spreads to the dentine, and even to the nerves and blood vessels in the middle of the tooth (see diagram on page 35).

The acid comes from tiny living things called bacteria. Everyone has millions of bacteria in their mouths. Normally they are quite harmless. But if sugary food gets stuck between the teeth, the sugar gives the bacteria energy. This is when the bacteria start to produce acid. The acid and the bacteria form a sticky covering on the teeth called plaque.

You can prevent plaque forming on your teeth by brushing regularly with a fluoride toothpaste. Fluoride helps to make the enamel on your teeth stronger. You can also use dental floss to remove any food that is stuck between your teeth.

If you do have tooth decay you may have to have a filling. The most common filling material for teeth is an alloy of mercury, silver, tin, copper and zinc. This is called mercury amalgam.

Growth and repair

All living things grow. Many plants grow for as long as they live. Humans grow until they become adults, and then the growth stops. Humans have two sets of teeth, but many animals grow new teeth throughout their lives. The teeth of a rat or a mouse replace themselves as they become worn down.

Many natural materials repair themselves if they are broken or worn-out. Bones grow back if they are broken. Some animals can even grow new limbs. For example, lizards often lose their tails. The tail is designed to break off easily if an animal grabs hold of it. Then the tail slowly grows back.

Many lizards have tails that are designed to break off easily. The tail will grow back slowly.

Key words
Bone a natural material that can repair itself.
Concrete a man-made material used for building roads and bridges.

Rubber and glue

Rubber and glue are examples of materials that are not usually used on their own to make things. Rubber, for example, is often used to make different parts for a car such as springy mounts for the engine. Glue is used to stick things together.

Rubber

In 1876 a British explorer smuggled 70,000 rubber seeds out of Brazil. Until that time, rubber trees only grew in South America.

▽ Tapping a rubber tree for latex.

Rubber is a natural polymer (see page 14). It is made from a white liquid called latex. Latex comes from plants, particularly from the rubber tree. People cut a strip of the bark from the tree. The latex drips from the tree into a cup. This process is called 'rubber tapping'.

Rubber is very elastic. It can be stretched, and then it will return to its original shape. In rubber, the long polymer chains (see page 14) are held together by weak bonds. This means that when the rubber is

△ To help rubber keep its shape, different materials are added to it. Tyres are made from rubber mixed with sulphur and other elements to make it harder.

EXPERIMENT

Rubber and spaghetti

In this experiment you will examine the way lengths of spaghetti slip over each other. You can imagine that the pieces of spaghetti are like the polymer chains in rubber.

You will need: 20 long pieces of spaghetti, a little oil, a large saucepan, a plate, two forks, a colander, some water and the use of a cooker.

Warning Take care when using the cooker, and ask an adult for help if you need it.

1. Fill the saucepan half full of water. Heat the water on the cooker. When the water is boiling, add the spaghetti. Cook for 7 minutes.
2. Drain the hot water using the colander.
3. Rinse the spaghetti in the colander with cold water. Add 2 teaspoons of oil and mix it in. Leave to cool.
4. After about 10 minutes, try to pull the lengths of spaghetti apart.
The oil prevents the lengths of spaghetti from sticking to each other, like the polymer chains in natural rubber.

 Why do racing cars use tyres made of soft rubber?

squeezed, the chains slide over one another easily. However, when the squeezing stops the bonds between the polymer chains are strong enough to make the chains spring back into place. When rubber is heated, the polymer chains move even more easily. This means that rubber can be poured into special shapes, or moulds.

Glue

Glue is a very important material in our everyday lives. Glue is often liquid and sticky before it is used. But it hardens to become a solid that holds objects together. There are both natural and man-made glues.

Mussels make a glue to attach themselves to rocks so that they are not washed away by the waves.

Many animals and plants make sticky materials. For example, the female spider makes a web of silk threads (see page 15). The outer part of the web is sticky to trap insects. The mussel attaches itself to rocks with a glue that it makes itself. Conifer trees make a sticky substance called resin.

EXPERIMENT

Making glue from milk

In this experiment you will make a natural glue from milk. This glue was used thousands of years ago by the Ancient Egyptians.

You will need: 500 cm³ of skimmed milk, a little baking soda, some vinegar, a saucepan, a glass bowl and an old spoon.

Warning: take care when using the cooker, and ask an adult for help if you need it.

1. Pour the milk into the saucepan and add 100 cm³ of vinegar. Heat the milk gently until lumps form.

2. Pour the milk into the bowl and let it cool. You should find a large lump of rubbery material at the bottom of the bowl, covered by a layer of watery liquid.

3. Pour off the watery liquid. Mix the solid material with 25 cm³ of water and a teaspoon of baking powder. A chemical reaction takes place which makes the glue.

The resin drips out of the bark of the tree in small, sticky drops.

There are two kinds of man-made glue, solvents and polymerisers. Solvents are usually liquids. A solvent glue is made from a solid that is dissolved in a solvent such as water. When you spread the glue, the molecules of solvent evaporate into the air, leaving the solid behind. The solid sticks the surfaces together.

Polymerising glues do not need solvents. A polymerising glue is made from a monomer (see page 14) and a hardener. When the monomer and hardener are mixed, they form a new polymer which becomes hard. As it hardens the new polymer sticks two surfaces together. The two parts of polymerising glues have to be kept separate until they are used. For this reason, these glues are usually sold in two tubes. Araldite and super-glue are both polymerising glues. These glues are very powerful and should be used with great care.

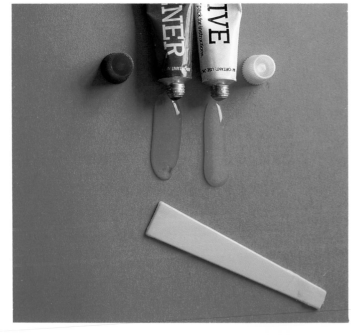

The two tubes and mixer of a powerful, man-made polymerising glue.

The sundew plant makes a natural glue to trap small insects.

Super-glue is so good at sticking to skin that doctors sometimes use it for attaching skin grafts.

Key words
Glue a material that joins together two surfaces.
Rubber an elastic material made from latex.
Solvent a liquid in which a substance will dissolve.

The future

We can be certain that people will develop many new materials in the future.

The latest kind of plastic can withstand heat up to a temperature of 2700°C. To show how effective this plastic is, an egg was coated with the plastic. Then the egg was held in a very hot flame. The egg came out of the flame uncooked and undamaged. This plastic will be useful for protecting furniture from fire, and even for protecting space vehicles.

A material of the future. SEAgel is so light that is can be supported by bubbles of air.

Balsa wood is a strong but very light wood. However, supplies of balsa wood are running out, so new materials are needed to replace it. A new invention is called Safe Emulsion Agar Gel or SEAgel. It is made from a kind of seaweed called kelp. SEAgel is a solid material but it is almost as light as air. Because it is made from kelp, it is biodegradable (see page 20). It is also safe to eat. It will probably be used for packaging, in medicines and in fridges.

Some of the most exciting new materials for the future are composite materials. These are materials that are made from more than one substance. For example, reinforced concrete (see page 32) is a composite material. Some cars already make use of composite materials such as fibre-reinforced metal alloys. These are metals that are reinforced with tough fibres. They help to make car parts lighter and stronger.

Many scientists are studying the natural world to learn more about materials. For example, scientists looked at the structure of the abalone, a mollusc with a thick shell. The shell is a composite material with pieces of calcium carbonate in a polymer, rather like cement joining chalk bricks. The design of the abalone shell is very strong. In fact, it is so strong that a similar design was used for new armour for tanks!

Many materials in the the natural world react to changes around them. Scientists are now designing man-made materials that will also react and change. These materials are called 'smart' materials. For example, 'smart' gels, called polymer gels, can move around rather like artificial muscles. In the future, there will also be 'smart' concrete. This concrete will be able to repair itself just like bones mend themselves. These developments show how important it is that scientists continue to study the natural world.

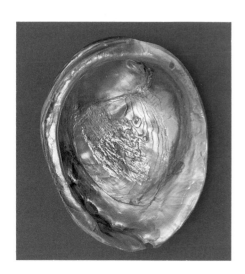

The shell of the abalone is very tough.

This photograph of Sydney, Australia shows both natural and man-made materials. Living things are made from materials that react to change. Most man-made materials cannot do this, although this may change in the future.

Glossary

Alloy a mixture of metals.

Atom the smallest part of a chemical element that can exist alone.

Biodegradable a substance that is able to be broken down naturally.

Bond an attraction between two atoms.

Bone a natural material that can repair itself.

Cellulose a natural polymer made from glucose.

Chemical reaction a change in which the atoms of two or more substances react to form new bonds.

Composite materials materials that are made from more than one substance.

Compound a substance made from two or more elements.

Concrete a man-made material used for building roads and bridges.

Corrosion a chemical reaction between air, water or another chemical on the surface of a metal.

Designer a person who plans the shape and style of something.

Dissolve when something dissolves into a liquid it becomes mixed with and absorbed into the liquid.

Element a substance made up of only one kind of atom.

Fibre a long, thin thread of material.

Gas a substance made up of atoms that can move around freely.

Gel a substance that is part solid and part liquid.

Glucose a kind of sugar.

Glue a material that joins together two surfaces.

Keratin the protein that makes up hair, nails, hooves and horn.

Liquid a substance made up of atoms that can move a little.

Lubricant a slippery material that makes two surfaces slide over each other easily.

Material the substance from which something is made.

Molecule a group of atoms bonded together.

Monomers the small parts that make up a polymer molecule.

Plastic a man-made material made from oil.

Polymer a large molecule made up of smaller parts.

Protein a natural polymer.

Rubber an elastic material made from latex.

Solid a substance in which the atoms are held together by strong bonds.

Solvent a liquid in which a substance will dissolve.

Thermoplastics plastic that can be melted by heating.

Thermosetting plastics plastics that do not soften when they are heated. Instead, they stay hard and in shape.

Index